Water Cascade: A Sequence

David H. Gibson

Water Cascade: A Sequence

Nazraeli Press

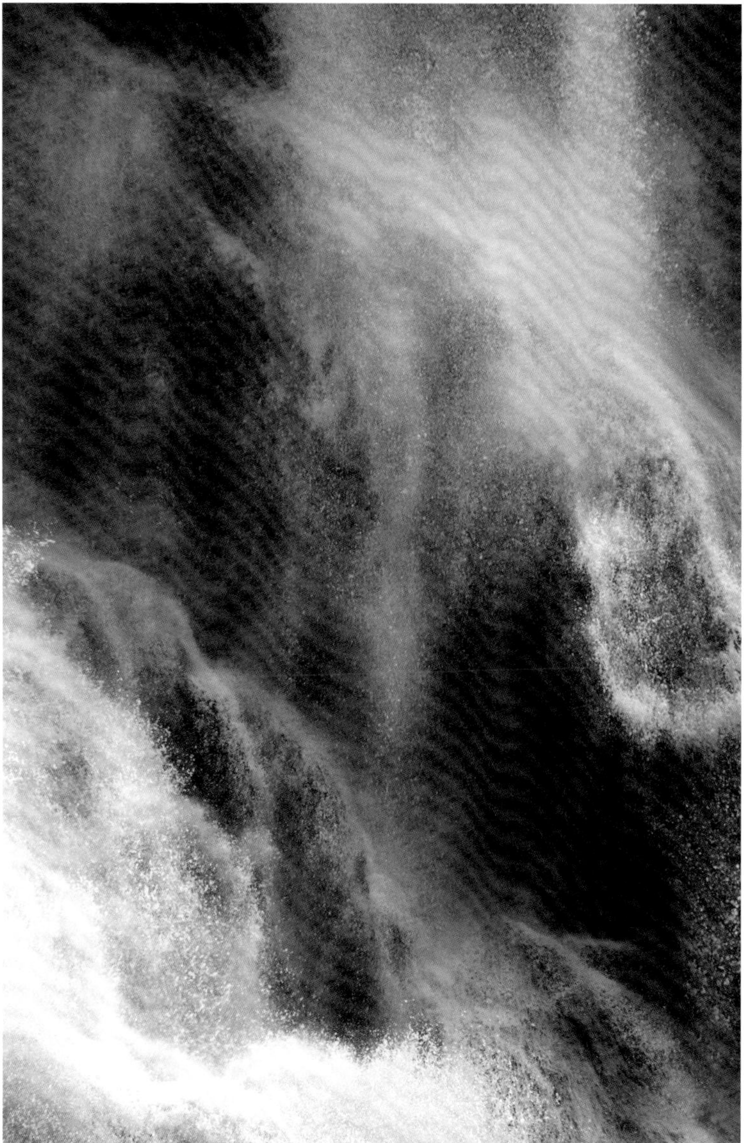

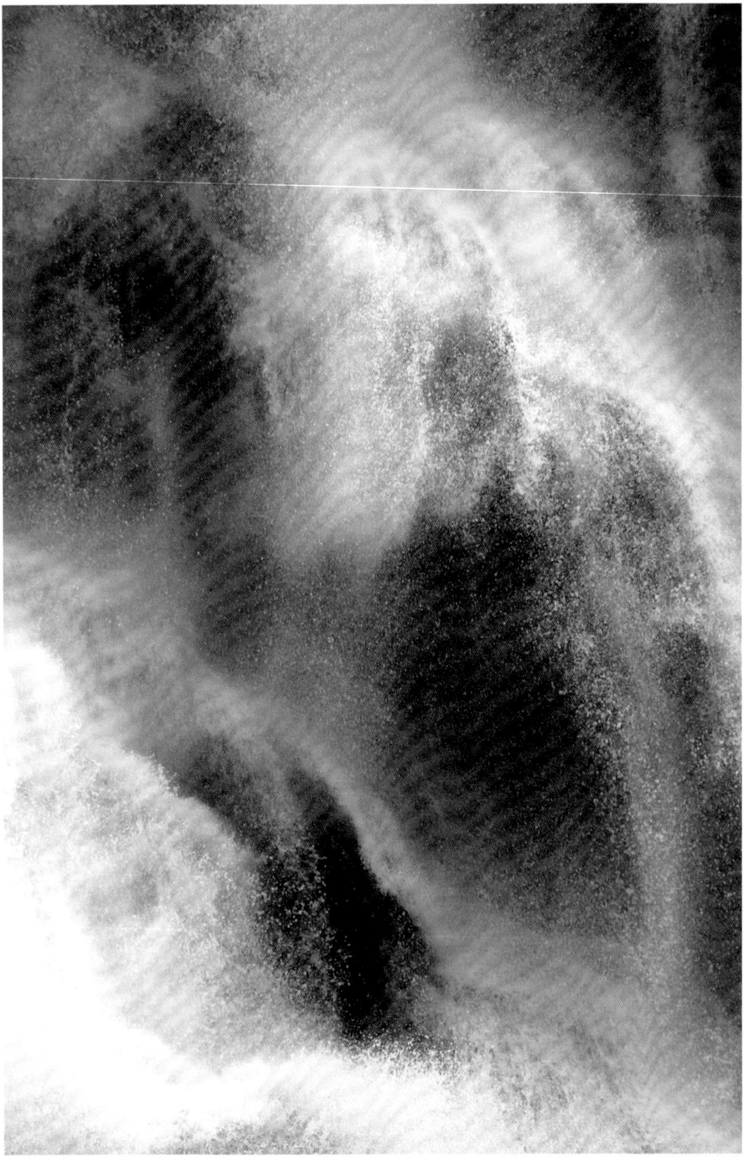

Water Cascade: A Sequence
A book of ten reproductions, and one original photograph
by David H. Gibson.

N° 423/ 500 (Print signed on verso)

Nazraeli Press One Picture Book N° 61
Printed in an edition of 500 copies

Order Address: Nazraeli Press
2871 NE Alameda Street
Portland, Oregon 97212, USA

Printed in China for Nazraeli Press
ISBN 978-1-59005-284-6